Collection: Tasmanian Museum and Art Gallery

A group of four Tasmanian tigers at Beaumaris Zoo, Hobart. These animals – an adult standing in the centre, and three young – were probably those caught by G. Wainwright at Woolnorth in 1914. This photograph was used in a publication by Beaumaris Zoo in 1915.

The Tasmanian Tiger in Pictures

Eric R. Guiler

St. Davids's Park Publishing

St. David's Park Publishing,
GPO Box 307C, Hobart, Tasmania, Australia 7001
1991

National Library of Australia
Cataloguing-in-Publication Data

Guiler, E. R. (Eric Rowland), 1922-
The Tasmanian tiger in pictures.

ISBN 0 7246 2209 8.

1. Thylacinus cynocephalus. 2. Thylacinus cynocephalus–Pictorial works. I. Title.

599.2

Design by MBA Publishing
Editing by Caroline Mordaunt

Printed and bound in Tasmania by Tasmanian Government Printing Office,
on fine paper produced in Tasmania.

INTRODUCTION

Tasmania is an island lying off the south-eastern tip of Australia and separated from it by Bass Strait. This strait was formed about 13,000 years ago when rising sea levels cut the island off from the mainland and Tasmania became isolated. One significant result of this was that some species of marsupials have survived in Tasmania although they have become extinct on the mainland.

MAP 1

▲ Sketch showing the position of Tasmania and the location of places on the mainland of Australia mentioned in the text.

The island is part of the Commonwealth of Australia and is about one-tenth the size of Japan. Much of the country is elevated land covered with thick forests with steep gullies making access difficult. The mountains are not particularly high, rarely more than 1,200 metres, and there are no permanent snow-fields. The rainfall is high in many areas, reaching more than 1,850 mm in some places. This makes farming impossible in these areas but native animals are able to live undisturbed where humans never go.

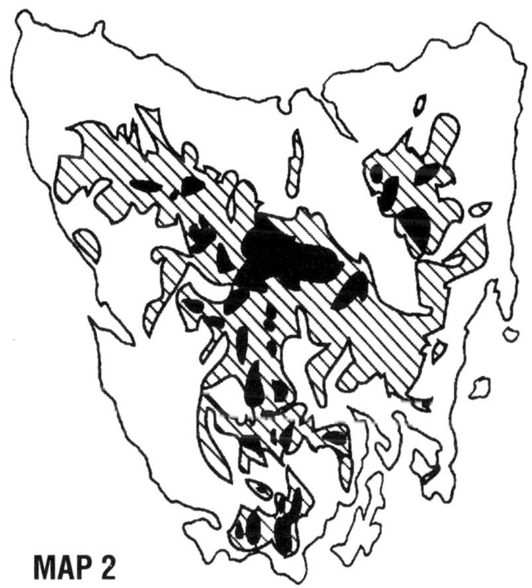

MAP 2

▲ Outline map of Tasmania showing the distribution of the high land.
Unshaded areas–less than 300 metres;
Shaded areas–300-1,000 metres;
Black areas–over 1,000 metres.

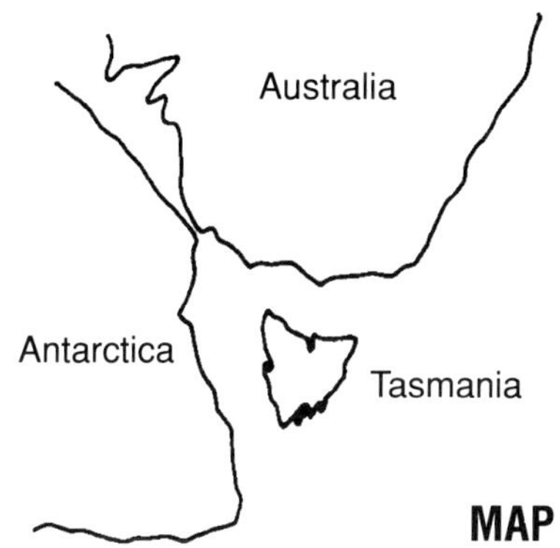

MAP 3

▲ The relationship of Tasmania to Australia and Antarctica during the break-up of Gondwanaland about 90 million years ago. Although separated from Australia at this time, it was connected at least eight times between then and the present day. This was due to fluctuations in sea level.

The Tasmanian tiger and the Tasmanian devil are both well-known nowadays in the scientific world for their survival stories, each of which has a different ending. The Tasmanian devil has survived the effects of changing habitat and food conditions and has increased greatly in numbers so that it is almost a pest in some places. It has been studied extensively from many aspects by scientists. The Tasmanian tiger story is very different and there is much speculation and mystery about this animal.

The last Tasmanian tiger in captivity anywhere in the world died in the Beaumaris Zoo in Hobart in 1936. Since then not many people have been lucky enough to catch a glimpse of it in the wild, and so far no animal has stayed within sight long enough for the viewer to take a photograph.

There are no photographs of the animal in its natural environment and few in zoos either and, although the female carried up to four pouch young, there are no pictures of the young in the pouch or playing with the mother. Indeed, we understand so little of the life of the Tasmanian tiger that it must be one of the least known large mammals in the world. Professor T. T. Flynn who came to the University of Tasmania in 1909 and is best known as the father of the actor Errol Flynn would have seen Tasmanian tigers in Beaumaris Zoo but he did not publish any information about them.

The Tasmanian tiger is a nocturnal animal and it was not often seen in daylight even when it was common last century. Occasionally it was recorded basking in the sun during the early morning. It was never easy to catch. Several thousands of them were killed and it is quite amazing that during this time there was an extraordinary lack of scientific interest in them. Few specimens exist in museums or other institutions and even photographs are rare.

The men who worked in the bush and trapped or observed the Tasmanian tiger and knew its habits are long-since dead and the stories we hear are third or fourth hand now and have been embellished greatly in the telling in most instances.

The Tasmanian tiger is a marsupial and so is related to the kangaroos, koalas, possums and all other marsupial fauna. However, its relationship to the kangaroo is not close, perhaps similar to that between the horse and the wolf.

It is a predator living on other species of marsupials, principally Bennett's wallaby and pademelon. We know little about any other food it may eat but it is reasonable to assume that it would eat any other warm-blooded animal such as the possum, and it may even eat lizards. It kills its prey, feeds off it and then never returns to the kill and it does not eat carrion. This would be of advantage in preventing it being poisoned whereas wild dogs can be destroyed in this way.

The Tasmanian tiger, known scientifically as *Thylacinus cynocephalus,* has several other common names. The most widely known are thylacine, Tasmanian wolf or hyaena. Obviously Tasmanian tiger and hyaena are derived from the stripes across the back and the base of the tail. The name thylacine comes from its generic description and is the name favoured by science but to the people of Tasmania it will always be known as the Tasmanian tiger or Tassy tiger.

This little book shows a selection of photographs most of which have not been published before. The text on the various expeditions is based mainly on my own experiences and the only scientific reference work is my own book *Thylacine - the tragedy of the Tasmanian tiger.* It is extraordinary that an animal which attracts such widespread interest has not been the subject of more detailed studies.

This collection of photographs has been gathered over the years. A few have appeared in various short articles. This is the first attempt to bring together a photographic record of the life of the Tasmanian tiger. The collection has many gaps. For example, as mentioned we have no pictures of the pouch, the young just after they leave the pouch, mating and so on. Perhaps this publication will produce a few 'new' pictures from some long-forgotten collection. The localities at which the photographs were taken are shown on Map 4 on page 8.

I am indebted to the Tasmanian Museum and Art Gallery and to the Archives Office of Tasmania for permission to use photographs from their collections. Many of the prints were taken from a short movie film now in the possession of the Department of Parks, Wildlife and Heritage.

▼ Male and female animals taken at Beaumaris Zoo, Hobart, in the early years of this century. Notice the size differences and the different shape of the head, particularly the jaw muscles. The Tasmanian tiger is unique in the marsupials in that both sexes have a pouch. The female carries its young in it and the male has a 'pseudopouch' to protect its testes.

Collection: Tasmanian Museum and Art Gallery

Collection: Archives Office of Tasmania

▲ A tiger photographed at Beaumaris Zoo, Hobart (date unknown).

The Australian Aborigines came to the continent from the north some 40,000 or more years ago. Presumably these people soon made contact with Tasmanian tigers as there is fossil evidence that they existed long before Man's time in Australia.

▶ We know that Tasmanian tigers continued to live in Australia until about 3,000 years ago because cave paintings dated to that period are found at various sites. This drawing is at Obiri Rocks in Arnhem Land (Map 1).

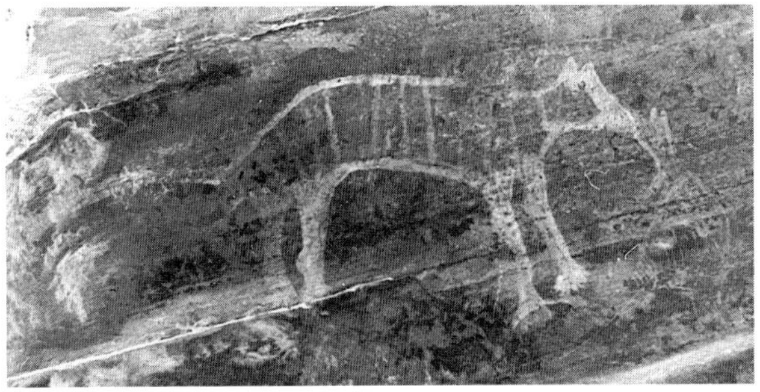

Author's Collection

The arrival of these people in Australia started a gradual decline for the Tasmanian tiger. The Aborigines acquired dogs which soon were running wild. An Australian native dog, the dingo, arrived at about the same time as the dogs. Both these species roamed over the continent and competed with the Tasmanian tiger for food and possibly may have fought with it also.

In Tasmania the situation was different as the rising waters of Bass Strait had cut Tasmania off from the mainland and so protected the Tasmanian tiger from conflict with these competitors.

The Tasmanian Aboriginal population was also isolated and it is recorded that they killed the Tasmanian tiger for food but probably they were never a serious predator.

At one time (1868) it was believed that there were two species of Tasmanian tiger, a 'greyhound' and a 'bulldog'; however, these turned out to be sex differences found in young animals.

▼ This is a male with a narrow head and snout. The male was sometimes called the 'greyhound' tiger.

Author's Collection

Author's Collection

▲ Here we have a female which has a wider head and snout. The pouch is under her belly and can carry up to four young during the first few months of their life.

The beginning of European migration to Tasmania in 1803 was the start of the invasion of the last stronghold of the Tasmanian tiger and was to be disastrous for the animal.

▶ The early settlers soon saw Tasmanian tigers but it was not until 1808 that the first scientific description was published. Some of the early artists never saw a Tasmanian tiger and used their

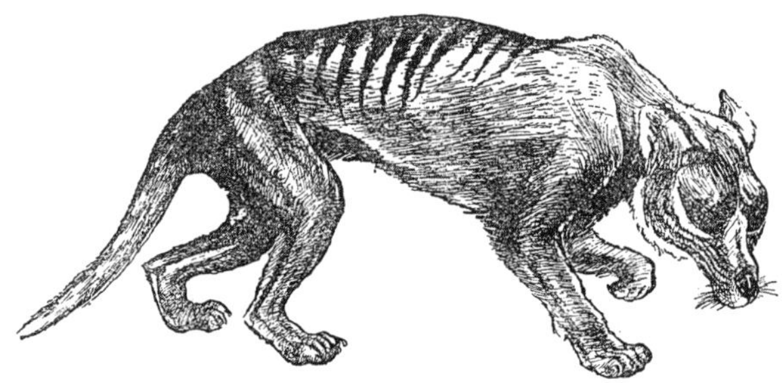

Collection: Archives Office of Tasmania

imaginations to produce some remarkably inaccurate drawings.

▼ However, this was soon corrected and more realistic drawings appeared. John Gould, the eminent wildlife engraver, in 1863 made a very good painting of the head of a Tasmanian tiger.

Tasmanian Marsupial Wolf.

Collection: Archives Office of Tasmania

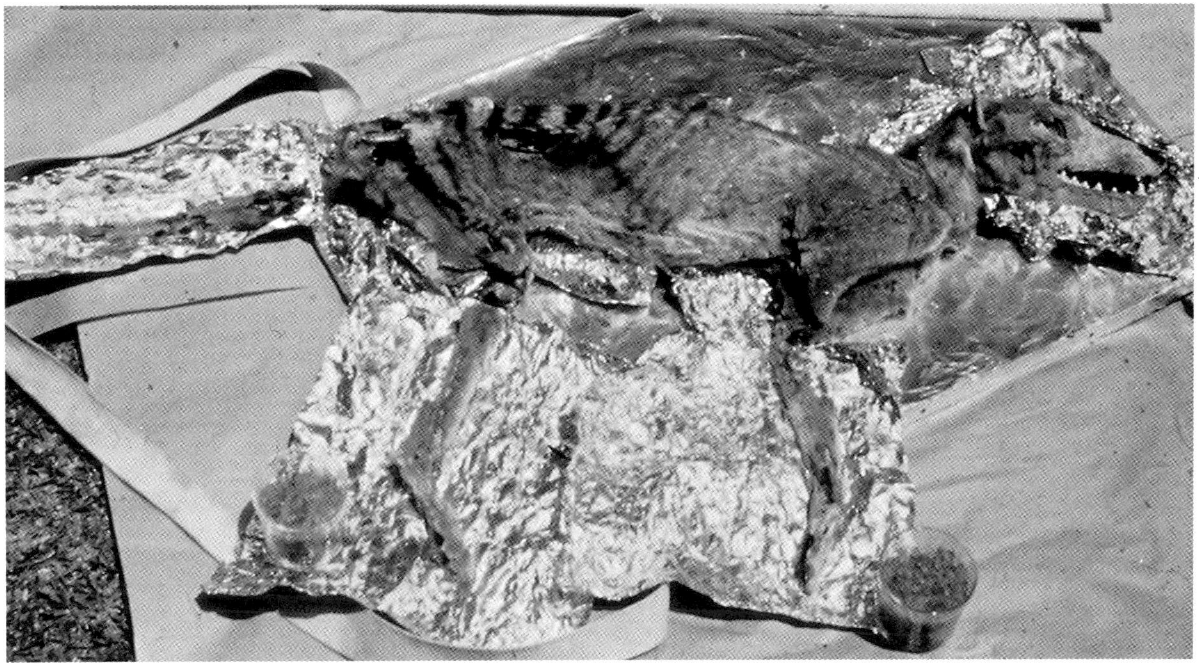

Collection: Western Australian Museum

▲ ▶ Startling evidence of the presence of Tasmanian tigers in Australia was found in a cave on the Nullarbor Plain (Map 1) when a mummified specimen was found. This unfortunate animal had fallen into the cave, dying there and was mummified in the dry atmosphere. This happened about 3,300 years ago and proves that the animal was living in semi-desert as well as in the wetter places such as Tasmania. However, there are no confirmed reports of a live Tasmanian tiger ever having been found in recent years living on the mainland.

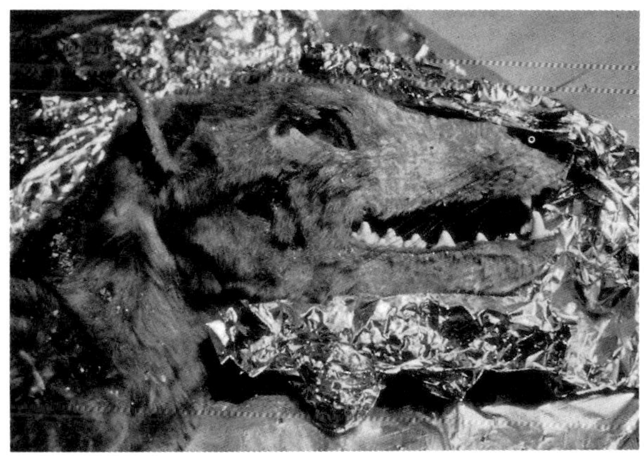

Collection: Western Australian Museum

HABITAT

Tasmanian tigers once lived all over Tasmania from the mountain tops to sea level. They preferred the open woodlands and coastal heath where there was abundant food for them to hunt. These coastal dunes at Studland Bay (Map 4) were a favoured habitat. ▼

Author's Collection

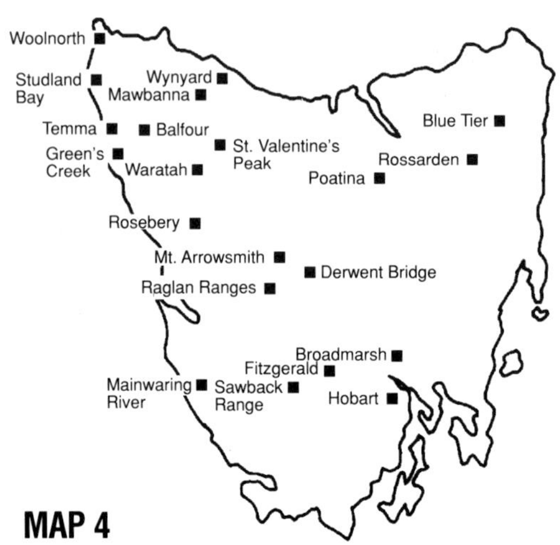

MAP 4

▶ Forests surrounding open grasslands have large wallaby populations which come out to the clearings to feed. Tasmanian tigers used to prowl the fringes of these forests to catch the wallabies. This photograph shows the Western Tiers near Poatina (Map 4) with forests surrounding the cleared grazing land.

▼ St. Valentine's Peak in north-west Tasmania (Map 4) shows some of the high mountain country. The silver tussock (a native grass) plain surrounded by thick forests would offer shelter and grazing to wallabies as well as cover to preying Tasmanian tigers.

Author's Collection

Author's Collection

Author's Collection

Author's Collection

▲ This country near Rosebery (Map 4) shows another type of western vegetation. It is very difficult to walk through these forests carrying camera gear or traps. The scrub in the foreground is head high and further back the prickly bushes are about four metres high and close together with vines intertwining them. Under the trees there is thick undergrowth while the ridges are invariably steep with cliffs or muddy slippery sides.

Author's Collection

▲ This mountainous area at the Blue Tier (Map 4) on the east coast was once renowned for its Tasmanian tigers. Nowadays there may be one or two left there.

◀ A typical west coast scene at Mt. Arrowsmith (Map 4) with bare rocky mountain tops and forested lower slopes. The button grass plain in the centre of the picture would provide limited food for some wildlife. These plains cover extensive areas in western Tasmania.

These areas are not used by Tasmanian tigers because they cannot move quickly enough through the undergrowth to catch their prey.

WOLF-LIKE FEATURES

In many ways it would be better to call the Tasmanian tiger by its other common name, the Tasmanian wolf, because it has so many wolf- or dog-like features. This similarity arises from both the wolf and the Tasmanian tiger having the same lifestyle in which they hunt their prey by running it down.

Tasmanian tigers do not have any cat-like features, but take their name from the stripes on their back.

▼ Notice the deep chest, slim belly and long legs which are typical of an animal which hunts by running down its prey. The head and ears are very like those of a dog or a wolf.

Author's Collection

▲ They even walk like a dog. See the way the left fore-foot has curled backwards as it walks.

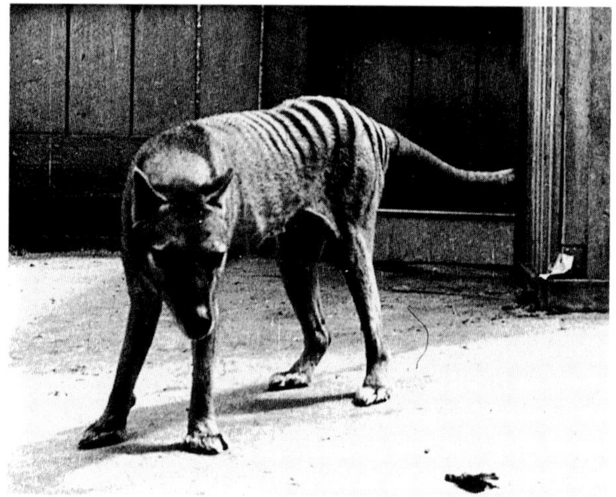

Author's Collection

▲ How many times has one seen a dog sniffing in doubt like this? ▼ They lie down like a dog.

Author's Collection

Author's Collection

Author's Collection

Collection: Archives Office of Tasmania

Author's Collection

◀ ▲ They sit like a dog. And they get up like a dog.

These similarities even extend to the animal's internal structures. The skeleton is very dog-like, emphasising the deep chest typical of an animal that runs after its prey.

▶ Even the skull is similar to that of a wolf. The Tasmanian tiger skull is on the right. It can be recognised by the presence of a small canal just in front of the orbit. This canal carries the duct from the tear gland to the orbit. The lachrymal canal is just inside the orbit in wolves and dogs.

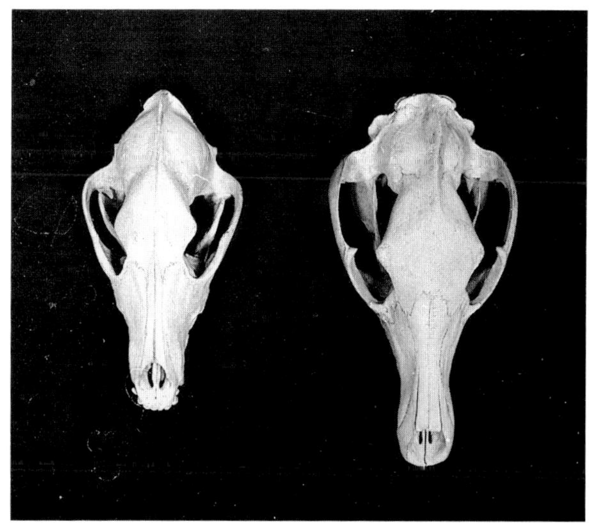

Collection: Archives Office of Tasmania

However, there are some features of the Tasmanian tiger which are not like those seen in the wolf.

▶ The Tasmanian tiger has a most spectacular gape, very much greater than that seen in wolves and dogs. This enables it to catch and hold large prey such as kangaroos.

In this it is assisted by a full set of very strong teeth which are of similar form to those of the wolf. However, the Tasmanian tiger has more teeth than a wolf having a total of 24 compared to 21 in the wolf. The teeth are arranged differently; the Tasmanian tiger has four upper incisors (three in the wolf), three upper premolars (four in the wolf), and four molars in both the upper and lower jaws (two and three respectively in the wolf).

Collection: Archives Office of Tasmania

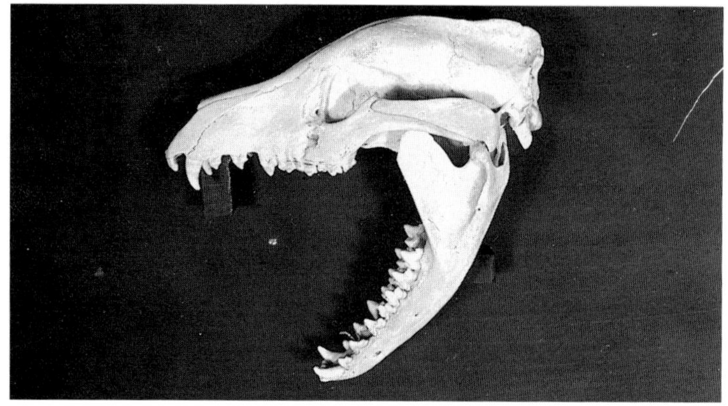

Author's Collection

▲ The skull shows clearly how the lower jaw is articulated in such a way as to allow the wide gape.

▶ This position with the rear feet and ankle joint flat on the ground is not used by dogs, Sometimes in tracks in the bush it is possible to see part of the very long hind foot pads in the mud and this makes identification of the spoor, or trail left by the animal, easy and certain.

▶ Having the use of large rear feet can be handy when standing up on two legs. This position is similar to that used by kangaroos when the hind feet and tail are used to maintain stance and balance. The tail of the Tasmanian tiger is semi-rigid and can be used as a support whereas that of a dog is floppy and bends both sideways and up-and-down and cannot be used for support.

▶▶ In common with dogs, the Tasmanian tiger uses its forepaws in manipulating food.

Author's Collection

Author's Collection

Collection: Tasmanian Museum and Art Gallery

4 PERSECUTION

The Aborigines in Tasmania were relatively few in number living in small tribes scattered around the island. They lived mainly by catching game and shellfish and had no agricultural or technical skills which could alter their environment. The new settlers soon upset this balanced state.

Land was cleared and farms established upon which domestic animals and crops were farmed. These had to be protected from native animals which ate them. As well as this the native mammals were killed for food and for their pelts.

In breaking up the land for farming and cutting down the forests the new settlers destroyed much native habitat.

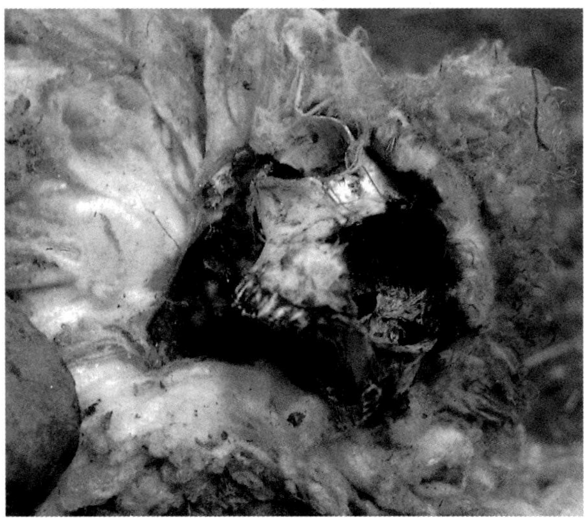

Author's Collection

Tasmanian tigers were distributed all over Tasmania although they were never numerous. Some of them turned to killing sheep and came into conflict with the farmers who mistakenly claimed that the Tasmanian tigers killed many thousands of their sheep. In fact, much of the killing was done by wild dogs which had escaped from the settlements and farms.

◀ Tasmanian tigers would kill sheep in a very characteristic way. They used their huge gape to bite out the throat and then they drank the blood. They ate the vascular nasal tissues followed by the liver, kidneys and some of the other meat. Dogs kill sheep differently, pulling their wool and worrying them to death. Tigers killed kangaroos by biting the throat and then ripping open the rib cage, eating the heart, lungs and liver while lapping up the blood. As mentioned, tigers would never return to a kill.

As early as 1830 the Van Diemen's Land Company, which raised sheep and cattle at Woolnorth in north-western Tasmania (Map 4), paid a reward for every Tasmanian tiger that was killed on its property.

Towards the end of last century it was believed that large numbers of sheep were still being killed by Tasmanian tigers and the government introduced a bounty scheme over all of Tasmania to try and reduce this predation. The reward was £1 per animal with 10 shillings per pup which was the equivalent of about a week's wages. This scheme was very successful and proved to be disastrous for the species so that by about 1910 the number of Tasmanian tigers had become critically low.

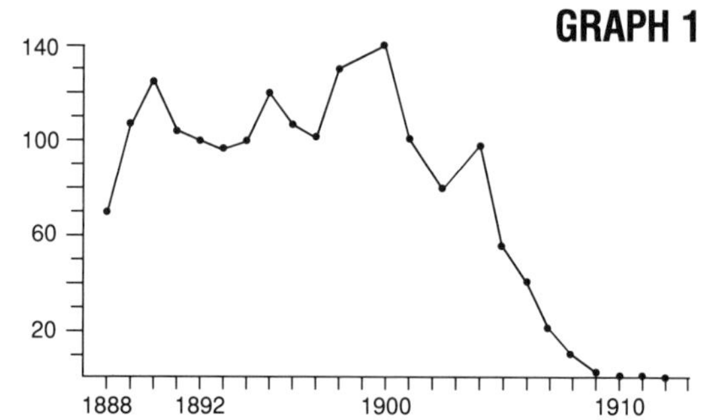

Graph 1 shows the catch of Tasmanian tigers during the official government bounty period, 1888-1910. The figures on the left show the number of Tasmanian tigers caught each year.

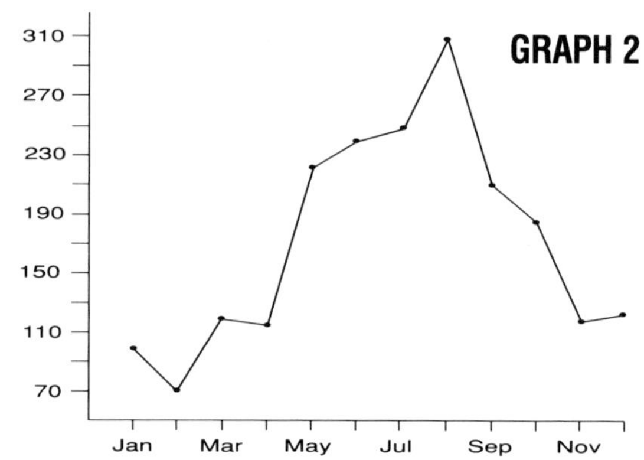

Graph 2 shows the catch for each month of the bounty period 1888-1910, based on various official records. Most tigers were caught in the winter. The cold weather and mountain snow may have forced them to lower country round the farms or, more likely, they were caught as part of a large trapping industry which took place in winter to catch fur-bearing animals.

Meanwhile the Van Diemen's Land Company was still catching tigers at Woolnorth. Although hunted continually on this property since 1830 the largest number caught in any one year was captured in 1900. Graph 3 shows the catch since 1874. This would suggest strongly that hunting was not the only factor involved in their decline.

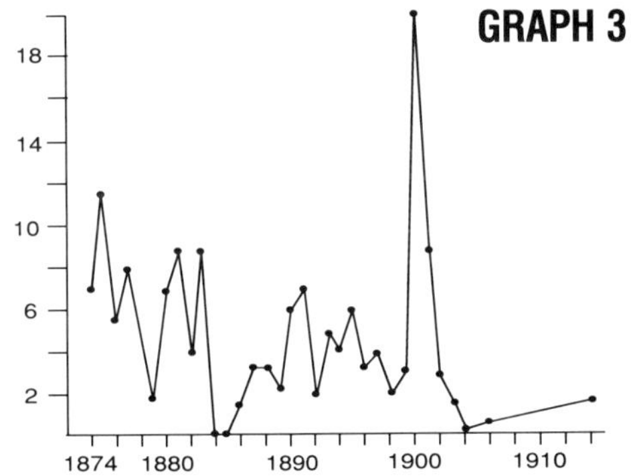

GRAPH 3

It is believed that another important reason for the drop in numbers of the Tasmanian tiger was a disease which swept through the remnants of the population during 1908 killing many tigers. This, together with hunting, habitat alteration and the fact that they never were numerous, was sufficient to bring them close to extermination.

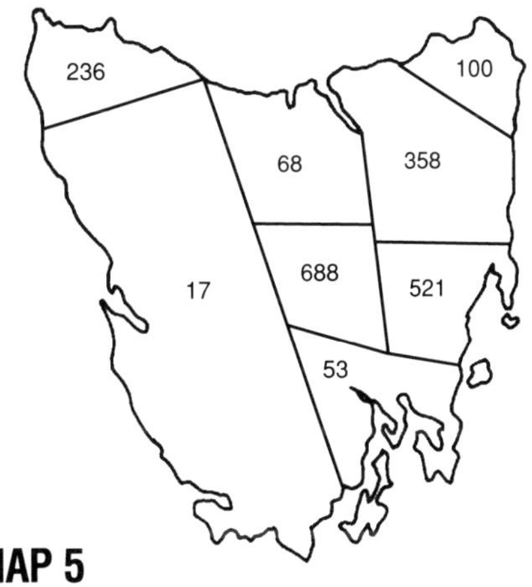

MAP 5

▲ This map shows the number of Tasmanian tigers caught for government reward in different parts of Tasmania during the bounty period 1888-1910.The largest number caught at any one place was at Derwent Bridge in the centre of Tasmania (Map 4) where 235 were killed.

Notice how the distribution follows the high land as shown on Map 2 and also observe the small numbers caught in the south and west. This is because there were not many people in those areas and there were few roads. There would not have been many Tasmanian tigers living in the dense forests of the region.

▼ A successful hunt for a Tasmanian tiger brought in some extra money which was welcome in hard times. This one was killed in 1922 near Waratah (Map 4) when it was hit on the head with a piece of wood.

Usually they were caught in a snare and either captured alive, shot or knocked on the head. It was not unusual for a trapped animal to die in the snare before being found by the trapper.

Author's Collection

Collection: Tasmanian Museum and Art Gallery

Even during times when Tasmanian tigers were relatively common it was quite an event to kill one. The successful hunter for a short time became a person of note in the community and photographs were taken of the trophy. When the animal became very rare people would travel miles to view one that had been killed or captured.

◀◀ This tiger was killed at Fitzgerald around 1912 by a Mr Quarrel and his dog Crisp. He was paid £5 for the carcase. The other man is 'Dang it all' Pearce.

◀ Many dead Tasmanian tigers were brought to this house at Derwent Bridge (Map 4). The hunter who lived here, H. Pearce, claimed the reward for 53 tigers, all of which were caught within 25 km. of this house. His family caught another 23 Tasmanian tigers.

▼ On the rare occasions that a Tasmanian tiger was sighted it provided an exciting hunt and was a change from the tedium of nineteenth-century farm life. The hotel at the small seaport of Whale's Head, now called Temma (Map 4), tried to make an attraction around 1910 of the sport although there were never many tigers in the area. The hotel no longer exists and there are no Tasmanian tigers to be seen either.

Author's Collection

SMITH AND CARTLEDGE

Remind their Friends and the Travelling Public that they

HAVE OPENED A

COMMODIOUS HOTEL

AT

WHALE'S HEAD,

The Port of the Balfour Copper Field, where Visitors will find nothing but the Best. Tourists in search of fun can be accommodated with Kangaroo Hunting, Tiger Shooting, Fishing, Yachting, &c., &c. Horses provided when necessary

COMMUNICATIONS PROMPTLY ATTENDED TO.

Collection: Archives Office of Tasmania

Collection: Archives Office of Tasmania

Collection: Archives Office of Tasmania

▲ How many Tasmanian tigers ended up like this? Their skins were pegged out on trees or the sides of buildings and fences to dry before being preserved.

It was said that many skins were sold to be made into men's waistcoats which were in fashion in England in the 1880's. The skins were sold in Hobart for about £3 18s each which was more than the basic wage at that time.

Most Tasmanian tigers were killed after they were captured but those that were taken alive sometimes died soon after being taken into captivity. A small number of animals were exported to zoos throughout the world but they were not popular as display animals. Being mainly nocturnal they spent most of the day asleep. Apparently in captivity they smelled more offensively than most other animals.

▼ The last known killing of a Tasmanian tiger took place at Mawbanna in 1930 when Wilf Batty shot one which was around his hen-house. Notice how reluctant Wilf's dog is to go near the animal. Dogs showed great fear of tigers when scenting them in the bush and would cower beside their owner when a tiger was in the vicinity.

Collection: Tasmanian Museum and Art Gallery

Author's Collection

▲ View from the Batty home site as it is today looking across cleared pastures to thick scrub in the background. The house and outbuildings have been extensively altered and there is no sign of the chicken shed where the Tasmanian tiger was killed. Dense forest and scrub originally covered this area but now remains only in the gullies and steep hillsides.

▼ Wilf Batty with Dr Elaine Murphy of N.Z. T.V. during the preparation of a documentary, 1989.

Courtesy: Bill Donovan

Collection: Archives Office of Tasmania

▲▶ Each of these two people ran successful export businesses of Tasmanian tigers during the 1890-1920 period. Mrs Roberts owned Beaumaris Zoo in Hobart and exported live animals overseas to zoos. Mr J. Harrison of Wynyard (Map 4) supplied Tasmanian tigers both within Tasmania to Mrs Roberts and to overseas and possibly exported their skins as well. A live Tasmanian tiger could be purchased for £8–£20 and re-sold to London Zoo for £30–£40 in 1910-1920.

▶ Beaumaris, Sandy Bay Road, Hobart, the site of Mrs Roberts' zoo where Tasmanian tigers were kept on display. Others were held there prior to export to overseas zoos. The house still exists but the gardens have been totally destroyed. This photo was taken *circa* 1940 by which time the zoo had been closed for 18 years.

Courtesy: Mr G. Roberts

Author's Collection

Many expeditions, some small scale, others lasting several years,have searched for the Tasmanian tiger. The first search was as early as 1936 when two police officers investigated in the Raglan Ranges (Map 4). They found footprints and droppings and concluded that some tigers still lived in the area.

The first of the long-term expeditions was organised by David Fleay in 1945-6. It operated for two years in several places near the localities searched by the police officers some ten years earlier. It obtained some footprints believed to have been made by a Tasmanian tiger.

The longest search was carried out by James Malley and J. Griffiths for four years from 1968-72. They investigated many remote areas over most of Tasmania and set up automatic cameras but did not obtain evidence of a Tasmanian tiger.

My own work started in 1958 and still continues but the search has not been continuous except in 1963-4 when a team consisting of myself and four others spent most of a year operating in the west and north-west.

We travelled over rough tracks and through creeks. Land Rovers were the only four-wheel drive vehicles readily available at that time. They were good cross-country motors but were not fast on the roads so travelling big distances took a long time. This vehicle we knew intimately as we dug it out of bogs, re-built collapsed bridges for it, nursed it over boulders and through sinking sands. It responded by being thoroughly reliable.

Author's Collection

▼ The 'main road' into Balfour (Map 4) in good condition. The track served as a river bed but we never got bogged. On one occasion we tried to avoid deep water by going across a grassy plain. The Rover bogged about 30 metres off the track and we spent three hours getting it out and on the track. After that we stayed on the 'road'.

Author's Collection

Author's Collection

Author's Collection

▲ Living in huts could be uncomfortable. This one at Green's Creek (Map 4) was a dreadful place. There was a five-centimetre gap between the concrete floor and the walls of the hut. The rain ran down the sides and the constant wind then blew it across the floor in streams. Needless to say, it rained every day.

▲ Searching the west coast at Cape Grim, Woolnorth, in 1964. The islands are The Doughboys. The tall man is Inspector G. J. Hanlon, B.E.M., who was seconded from the police to the fauna authorities for fauna policing duties. The author is standing beside him.

Author's Collection

◀ Living in tents was quite comfortable, especially in fine weather. This camp at Woolnorth was dry in wet weather and was cool in summer.

The big disadvantages of camping, in huts or tents, are wind, sleet or snow, possums, blowflies and Tasmanian devils. We knew of these hazards but sometimes a new one would appear. For example, on one trip in hot weather we put our butter in a small creek to keep it cool. When we returned we found that black jays had retrieved the butter from the creek and scattered it all over the place.

Author's Collection

▲ We had accidents, most were not serious. This Land Rover rolled down a bank between Temma and Balfour. Nobody was hurt but we lost several days' work getting it back on its feet again. It never was the same again and it was replaced by a Toyota.

Author's Collection

▲ Sometimes we found gardens abandoned whenever a farm was vacated. This one was at Balfour where there once was a copper mine. Inspector Hanlon (right) with Wildlife Officer R. Hooper amongst the hollyhocks.

We tried to snare a Tasmanian tiger during one of our expeditions. This was before we were aware of the hazards of post-capture trauma. We would never dream of such an approach nowadays.

▶ Snaring is a highly skilled profession. The objective is to catch the animal without killing it by entangling it in a noose of hemp. This is done by driving a stick into the ground and a treadle stick is placed across the animal trail. This stick has a branch coming off it so it can be held in position against the vertical peg. Both the peg and the stick each have a notch cut into them. A small piece of wood, the button, is shaped to fit into these notches.

Then a strong sapling is forced into the ground and then bent over and attached to the button in such a way as to raise the treadle about two centimetres off the ground. Also attached to the sapling is a loop of hemp which surrounds the treadle. The whole thing is then camouflaged with leaves.

When the treadle is tramped on by an animal it releases the button which in turn allows the sapling to spring into the air so causing the loop of hemp to tighten around the animal's leg and immobilise it. There it stays until we come around in the morning.

Author's Collection

Author's Collection

▲ Wildlife Officer K. Harmon preparing some of the sticks for making snares during our work at Green's Creek.

Author's Collection

Author's Collection

◀ Inspector Hanlon setting a snare at Studland Bay. We set 72,000 snares over six months catching plenty of game but no Tasmanian tigers. In fact we caught specimens of all the larger Tasmanian fauna. Occasionally we caught one of our team much to the amusement of the rest of us.

◀ Preparing some of our catch for the pot at Woolnorth. We were a long way from a butcher's shop so we ate some of the wallabies that we caught in our snares. Hindquarters are suspended on the bough of a nearby tree out of reach of any inquisitive dog, cat or Tasmanian devil. Author on the left with Inspector Hanlon.

▼ We tried to catch a Tasmanian tiger in a trap with an entrance wide enough not to harm the animal. Again we had no luck. We used bullock liver as bait, even dragging it for miles through the bush hoping to lure the tigers.

Author's Collection

▶ The trap was heavy and needed four men to carry it. Because of its size it was not possible to take it into thick bush. This trap was transported into the scrub on a bulldozer and then manhandled into position at Rossarden.

A later model trap was made of slot-together pieces which could be more easily carried. It had a central chamber which was proof against predators. We put hens in this as attractors. All we ever got was some eggs.

These traps were not successful being too cumbersome to shift around and they did not seem to attract animals.

Author's Collection

Author's Collection

▲ ▶ Two places near Rossarden (Map 4) where a Tasmanian tiger was seen in the last 30 years. This is open woodland country with deep gullies and some low overhanging cliffs which provide good shelter for all sorts of animals.

Author's Collection

Author's Collection

◀ Tigers were known to live in caves, hollow logs and thick vegetation. They apparently used a number of sleeping places in their home range.

▼ We found what we believed to have been a lair used by a Tasmanian tiger at Woolnorth in 1965 but its owner was not at home. The droppings were of the right size for those of a Tasmanian tiger and large but indistinct footprints could be seen in the dust.

Author's Collection

Author's Collection

▲ We used and still are using automatic cameras to take photographs of trail activity in places where we believe it possible that a Tasmanian tiger would walk in front of the camera. This camera was set on a track in the Sawback Range (Map 4).

▼ Sometimes we used helicopters to carry our gear into remote inaccessible areas such as the Mainwaring River (Map 4). We photographed most of the larger species of Tasmanian mammals.

Author's Collection

RECENT EVIDENCE

Author's Collection

What evidence do we look for when searching for a Tasmanian tiger? In the first place we interview any of the local residents who may have seen the animal. We then try to ascertain where they saw it so that we can search for footprints or droppings. We also search for other places in the area where we consider it likely to find evidence.

▲ Tasmanian tiger dung is very difficult to identify from the droppings of a dog or even a large Tasmanian devil. It may be possible to identify the dung by careful microscopic examination of the hairs contained in it. This is done in the hope that the animal has groomed itself and swallowed some of its own hairs. The task is so time consuming because the dung is usually full of the hair of the prey. This dropping at Studland Bay may be that of a Tasmanian tiger.

▼ We have spent a lot of time walking through the bush in likely places looking for footprints. These are drawings of the footprints of the left front foot of a Tasmanian tiger (1); the left foot of a dog (rear foot similar) (2); the left rear foot of a Tasmanian tiger (3). The impression of the ankle joint is rarely seen.

The footprint resembles that of a dog but the arrangement of the toes and claws in relation to the main pad is different. There are two clefts in the rear of the pad which are not found in the dog prints.

The problem is that it is difficult to find a good clear footprint; often the ground is too sandy, too hard or too muddy to retain the impression. Also, the shape of the print can change as the ground dries out, or is washed away.

We have taken many casts of footprints, some of which were of impressions made by Tasmanian tigers, others by dogs or wombats and a great many of unknown origin.

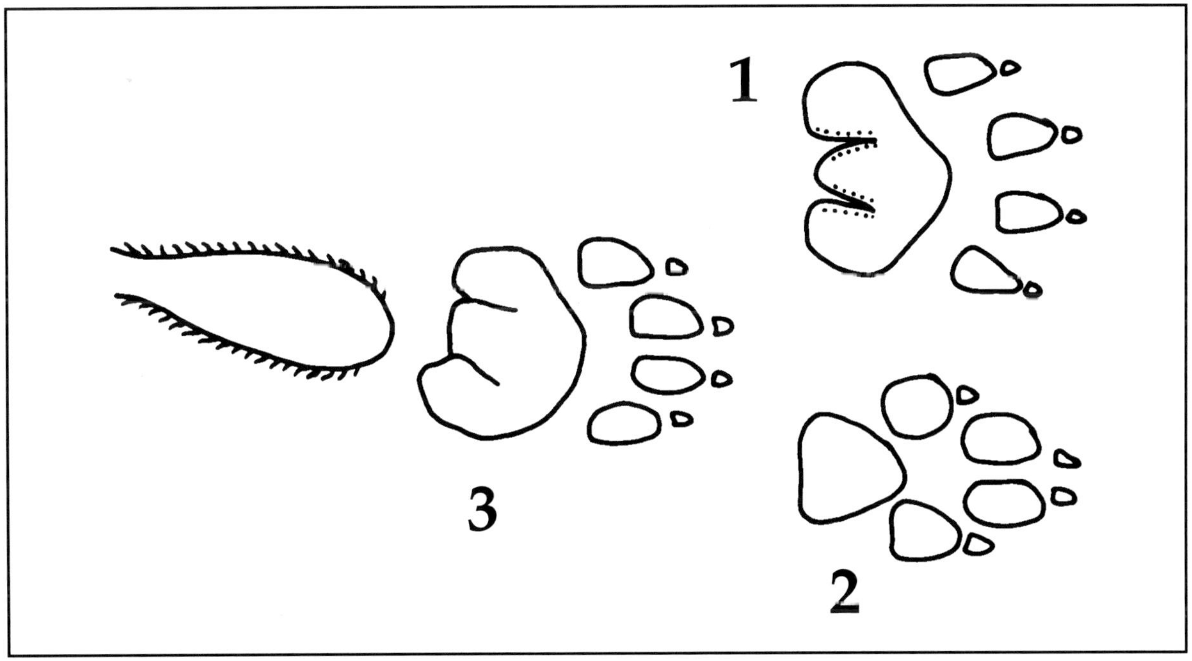

▶ The length of stride is important. If it is 69 cms. or more the prints could only have been made by a dog or a Tasmanian tiger. We made plaster casts of the prints for reference later. This series was found near our hut at Woolnorth in 1962.

Plaster casts are easy to make, especially if dental plaster is used as this sets quickly, even in water. Thus it can be used in rain or very wet tracks.

Many people write or telephone to say they have seen a Tasmanian tiger. We do not consider these reports to be reliable unless they have occurred in a place where other viewers, totally unknown to the reporter, have sighted a Tasmanian tiger. If a group of people claim to have seen a Tasmanian tiger at a short distance for a time of 20-30 seconds then the report is treated on its merits and slotted into the appropriate record.

▶ Map 6 shows the distribution of possible Tasmanian tiger sightings between 1936 and 1985.

▶ This map has the same divisions as Map 5. There are many reports from the west of Tasmania. This is in contrast to the distribution of catches shown on Map 5 when few were captured there. The reason for this is probably that there are many more roads through this area nowadays and therefore more people moving around in it.

Author's Collection

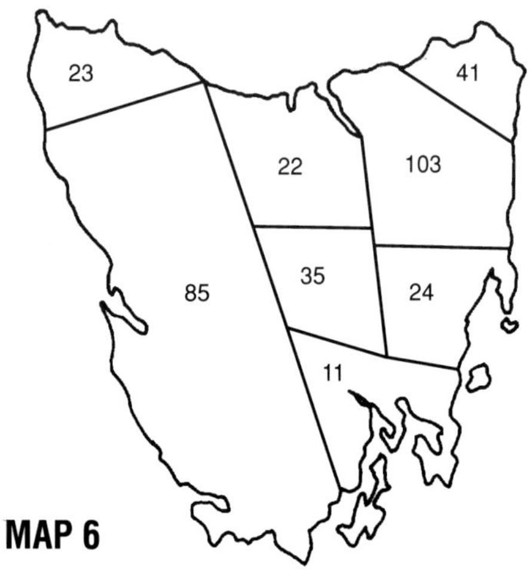

MAP 6

Later sightings are not revealed because of the dangers of people invading the area. It is imperiative that the few remaining Tasmanian tigers are left in peace in the hope that they may survive and even increase in numbers. We do not want tourists, photographers, 'do-gooders' and the curious stamping around the area. Nor do we want shooters as there are still, believe it or not, a few around who would shoot the last Tasmanian tiger.

In the almost total absence of any knowledge of the biology of the animal it is all that we can do.